Paulo Carteri Coradi

Drying and storing sunflower seeds

Paulo Carteri Coradi

Drying and storing sunflower seeds

Post-harvesting sunflower seeds

ScienciaScripts

Imprint
Any brand names and product names mentioned in this book are subject to trademark, brand or patent protection and are trademarks or registered trademarks of their respective holders. The use of brand names, product names, common names, trade names, product descriptions etc. even without a particular marking in this work is in no way to be construed to mean that such names may be regarded as unrestricted in respect of trademark and brand protection legislation and could thus be used by anyone.

Cover image: www.ingimage.com

This book is a translation from the original published under ISBN 978-613-9-62226-9.

Publisher:
Sciencia Scripts
is a trademark of
Dodo Books Indian Ocean Ltd. and OmniScriptum S.R.L publishing group

120 High Road, East Finchley, London, N2 9ED, United Kingdom
Str. Armeneasca 28/1, office 1, Chisinau MD-2012, Republic of Moldova, Europe
Printed at: see last page
ISBN: 978-620-7-71890-0

SUMMARY

AUTHOR

Prof. Dr. Paulo Carteri Coradi

He has a degree in Agricultural Engineering from the Federal University of Pelotas (2005), a Master's degree in Agricultural Engineering (Processing of Agricultural Products) from the Federal University of Lavras (2006) and a PhD in Agricultural Engineering (Processing of Agricultural Products) from the Federal University of Viçosa (2010) with a Sandwich period (SWE-CNPq) (2009-2010) at Kansas State University (Department of Grain Science & Industry), Kansas, USA. Post-Doctorate in Post-Harvest Engineering of Agricultural Products from the Federal University of Viçosa (2012). He was a PET, CAPES and CNPq scholarship holder during his undergraduate and postgraduate studies. He is a CNPq Productivity Fellow - Level 2. He is registered with the Regional Council of Engineering, Architecture and Agronomy (CREA- MG91048D). He is qualified by RABQSA International: Personnel and Training Certification as lead auditor for the ISO 9000 Quality Management System. He is qualified by Centreinar as a technical auditor of grain storage units and by MAPA as an official classifier of products of plant origin: qualification in corn, soybeans and sorghum. He was an Adjunct Professor at the Federal University of Mato Grosso do Sul (UFMS), Chapadao do Sul Campus (CPCS) (2011-2015). He is a member of the editorial board of the Brazilian Journal of Agricultural and Environmental Engineering (Agriambi). He is a reviewer for scientific journals, including: Drying Technology, Coffee Science, Revista Brasileira de Engenharia Agricola e Ambiental, Revista Engenharia na Agricultura, Revista Engenharia Agricola, Food and Bioproducts Processing, Revista Brasileira de Engenharia de Biossistemas, African Journal of Food Science, Ciência Rural, Journal of theScience of Food and Agriculture, Biomass & Bioenergy, Acta Scientiarum, Semina, Horticultura Brasileira, International Journal of Agricultural and Biological Engineering, Journal of Stored Products Research, Transactions of the ASABE, Food Security, Biosystems Engineering, Animal Feed Science and Technology. Since 2013, he has been a Permanent Professor in the Postgraduate Program in Agronomy (UFMS-CPCS). He is currently Adjunct Professor IV at the Federal University of Santa Maria (UFSM), Cachoeira do Sul Campus (CS), Coordinator of the Undergraduate Course in Agricultural Engineering and acts as a Permanent Professor in the Postgraduate Program in Agricultural Engineering (PPGEA) at UFSM. He has experience in the field of Agricultural Engineering, with an emphasis on Agricultural Product Processing Engineering, Agricultural Product Engineering and Post-Harvest Technology, working mainly on the following subjects: Processing, Drying, Storage and Quality of Agricultural Products.

Carlos Henrique Portela Fernandes

Agronomist graduated from the Federal University of Mato Grosso do Sul, Campus Chapadao do

Sul (2015).

Jean Carlos Helmich

Agronomist graduated from the Federal University of Mato Grosso do Sul, Campus Chapadao do Sul (2015).

ACKNOWLEDGMENTS

The Federal University of Santa Maria (UFSM) and the Postgraduate Program in Agricultural Engineering (PPGEA)

The Federal University of Mato Grosso do Sul (UFMS) and the Graduate Program in Plant Production.

CAPES, CNPq, FUNDECT and FAPERGS for their financial support.

CHAPTER I

DRYING KINETICS, MATHEMATICAL MODELING AND VOLUMETRIC CONTRACTION OF SUNFLOWER SEEDS (*HELIANTHUS ANNUUS L.)*

Summary: Given the prominence of the sunflower crop on the national scene and the importance of the drying stage for post-harvest, this work aimed to evaluate the drying kinetics of sunflower seeds in an oven with convective and forced air circulation at different temperatures, to describe mathematically and to determine the effects of drying on the volumetric contraction of the seeds. The experimental design used was entirely randomized, factorial (4x3), with four drying air temperatures (45, 55, 65 and 75 °C) and three replications. The seeds were dried until their water content reached hygroscopic equilibrium. Temperature and air flow were monitored using a psychrometer and a hot wire anemometer, respectively. The results showed that the drying time was significantly ($P<0.05$) reduced as the temperature of the drying air increased. The Wang and Sing model was the one that best fitted the experimental drying data. The effective diffusivity of sunflower seeds ranged from 2.83 to 2.93×10^{-6} m^2 s^{-1} , while the isosteric heat of sorption was 3.725.176 to 3.773.082 kJ kg^{-} 1. The reduction in water content influenced the volumetric contraction of sunflower seeds, especially at temperatures above 55 °C. The model by Correa et al. (2002) was the one that best fitted the experimental data on the volumetric contraction of sunflower seeds.

Keywords: optimization, post-harvest, dryer.

INTRODUCTION

Sunflower (*Helianthus annuus* L.) is the fifth largest seed crop and the fourth largest oil crop in the world (PESAGRO, 2007). Sunflower oil is one of the first oils to meet the quality standards demanded by the European biodiesel market, and has gained great commercial value due to the presence of special compounds that characterize it as a functional food. After harvesting, the seeds are subjected to pre-processing steps to guarantee quality in the subsequent seed manufacturing processes. Drying is an important post-harvest process. By reducing the water content of the seeds, they can be stored for longer, increasing marketability and product quality. However, drying processes need to be carried out carefully due to the transformations, deterioration and losses in the process, with the simultaneous transfer of heat and mass (CORADI et al., 2014a).

For proper drying, it is important to know the temperature of the drying air and the mass of seeds, the relative humidity and the air flow in the dryer. Carelessness in the drying process can damage the cellular structures of the seeds, denature the proteins and cause cracks and changes in color, reducing the physical and physiological quality of the seeds (ALMEIDA et al, 2009; OLIVA ET

AL., 2012; CORADI et al., 2014a). However, the drying process is lacking in sunflower seed efficiency studies. The method used today is more natural, in the open air, where the time required for drying is generally long and does not always result in a good quality product. Thus, natural drying is giving way to artificial drying techniques in order to increase the quality of the stored product (SILVA, 2008). In Brazil, there is little information on the artificial drying of sunflower seeds, such as drying kinetics, physical characterization and isothermal equilibrium moisture content, so the optimum drying conditions for domestic varieties are not well established (SILVA, 2008; SILVA et al., 2013; CORADI et al., 2014a).

The physical properties of agricultural products influence the optimization of industrial projects and the sizing of equipment used in harvesting operations and post-harvest processes (RESENDE et al., 2008). Knowledge of the physical and mechanical properties of agricultural products is of fundamental importance for the preservation of grains and for the construction and operation of various pieces of equipment used in the main post-harvest operations (MIR et al., 2013).

One of the main changes in the physical properties of agricultural products during the drying process is to reduce their external volume, i.e. shrinkage. The loss of water causes damage to the cellular structure of the product, leading to a change in shape and a reduction in the size of the tissue (MAYOR; SERENO, 2004; GONELI et al., 2011).

It is crucial to simulate and obtain theoretical information on the behavior of each product during water removal in order to develop and improve drying equipment. Mathematical models implemented in computer programs have shown significant results in non-linear and complex systems in various areas, including model control theory, recognition and decision analysis. These programs are able to solve problems that classical models, as a rule, are unable to do. They are called intelligent systems, among which Neural Networks and Fuzzy Logic stand out (CANEPPELE et al., 2010; CANEPPELE; SERAPHIM, 2010; 2013; SIQUEIRA et al., 2014).

In this study, this principle was applied to mathematical models that satisfactorily represent water loss during drying. The aim of this work was to study the drying kinetics of sunflower seeds in an oven with forced air circulation at different temperatures (45, 55, 65 and 75 °C), to fit the experimental data to non-linear mathematical models and to evaluate the effects of drying on seed shrinkage.

MATERIALS AND METHODS

The experiment was carried out at the Federal University of Mato Grosso do Sul (UFMS), Chapadao do Sul Campus (CPCS) in 2013. The sunflower seeds were collected at random and the impurities and damaged seeds were separated manually. The seeds were then dried in a convection oven with forced ventilation at temperatures of 45, 55, 65 and 75 °C. The seeds were dried until they reached hygroscopic moisture equilibrium. Three tests were carried out for each drying air temperature and 5 kg of sunflower seeds were used for each test. The temperature and relative humidity of the air were monitored throughout the drying process using a psychrometer. The water content (% b.u.) was determined by weighing 15 g of the sample. The samples were then placed in an oven with heating and air ventilation set at 103 °C ±1 °C for 24 h, according to the recommendations of BRASIL (2009). After this, the samples were removed and placed in desiccators. The water content (% b.u.) was determined by the difference in mass between the initial and final weight of the sample. The tests were carried out in three replicates.

The drying curves were fitted to the experimental data using thirteen different semi-empirical and empirical equations (equations 1 to 11), listed below:

Equations

$RX = \exp(-k - t)$ Newton (1)

$RX = \exp(-k - t^n)$ Page (2)

$RX = \exp(-(k - t)^n)$ Page Modified (3)

$RX = a - \exp(-k - t)$ Henderson & Pabis (4)

$RX = a - \exp(-k - t) +$ cLogarithmic (5)

$RX = a \cdot \exp(-k_o \cdot t) + b \cdot \exp(-k_1 \cdot t)$ Two Terms (6)

$RX = a \cdot \exp(-k \cdot t) + (1 - a) \exp(-k \cdot a \cdot t)$ Two-term exponential (7)

$RX = 1 + a.t + b.t^2$ Wang& Singh
(8)

$RX = a \cdot \exp(-k \cdot t) + b. \exp(-k .t) + \quad c. \exp(-k \quad .t)$ Henderson & Pabis (9)

$RX = a \cdot \exp(-k \cdot t^n) + b \cdot t$ Midili (10)

$RX = a \cdot \exp(-k \cdot t) + (1 - a). \exp(-k.b.t)$Diffusion approximation (11)

in which,

RX - humidity ratio, admensional;

t - drying time, h;

k, ko, k1 - drying constants, h^{-1};

a, b, c, n - model coefficients.

To determine the moisture ratios during drying under different conditions, the following expression (equation 12) was used:

$$RX = \frac{U - Ue}{Ui - Ue} \quad (12)$$

in which,

It is common to consider the value of the diffusion coefficient to be constant or linearly dependent on the air drying temperature. This relationship has been expressed by the Arrhenius model (MOHAPATRA & RAO, 2005).

$$D = A\exp\left(-\frac{E}{RT}\right) \quad (13)$$

in which,

A - constant $(m^2\ s)^{-1}$

E - activation energy $(kJ\ kmol)^{-1}$

R - universal gas constant $(8.314\ kJ\ kmol^{-1}\ K)^{-1}$

Tabs - absolute temperature (K).

The coefficients of the Arrhenius expression were linearized by applying the logarithm:

$$\mathrm{LnD} = LnA - \frac{E}{RT}\frac{1}{Ta} \quad (14)$$

The values for water activity, temperature and hygroscopic equilibrium humidity were obtained from the desorption isotherms of sunflower seeds, using the model that best fits the experimental data. To calculate the integral isosteric heat of desorption, we used the following equation:

$$Q_{st} = q_{st} + L = a \cdot \exp(-b \cdot U_e^*) + L \quad (15)$$

in which,

Q_{st} - integral isosteric heat of sorption $(kJ\ kg)^{-1}$

L - latent heat of vaporization of free water $(kJ\ kg)^{-1}$

Ue* - equilibrium moisture content (b.s.)

a, b - model coefficients.

The latent heat of vaporization of free water was obtained using the average temperature using the following equation:

$$L = 2502,2 - 2,39.T_m \quad (16)$$

in which,

L - latent heat of vaporization of free water $(kJ\ kg)^{-1}$

T m - average temperature (°C)

The unit volume contraction (Ψg) during drying was determined by the ratio between the final volume for each water content and the initial volume of the seed. The volume (Vg) of each seed was obtained during the drying process using a caliper according to the expression proposed by Mohsenin (1986):

$$V_g = \frac{\pi\ a\ b\ c}{6} \quad (17)$$

in which,

a - major axis of the seed (mm)

b - average seed axis (mm)

c - minor axis of the seed (mm)

Model references	Models	
Bala and Woods	$\Psi_g = a\{1-\exp[b(U-U_0)]$	(18)
Lang and Sokhansanj	$\Psi_g = a+\beta_1 (U-U)_0$	(19)
Rahman	$\Psi_g = a+\beta_2 (U-U)_0$	(20)
Corrêa et al.	$\Psi_g = 1/[a+b \exp(U)]$	(21)
Line	$\Psi_g = a+bU$	(22)
Exponential	$\Psi_g = a \exp(bU)$	(23)

in which,

Ψ_g - unit volume contraction (decimal)

U - water content of the product, in decimal (b.s.)

U_0- initial water content of the product, in decimal (b.s.)

β_1 - a + b (UR) + c (T)

a, b, c - parameters that depend on the product

RH - relative humidity (decimal)

T - air temperature (°C)

β_2 - volumetric coefficient, dimensionless contraction.

The experimental design used was completely randomized design (CRD) with three tests for each relative humidity and drying air temperature. Non-linear regressions were carried out to adjust the mathematical model analysis based on the QuasiNewton method, using the computer program Statistica 7.0® . To check the degree of fit of each model, the significance of the regression coefficient was determined by the t-test, adopting a 5% probability level, the magnitude of the coefficient of determination (R^2), the mean values of the relative error (P), the estimated mean error (SE) and the behavior of the distribution of the residuals were checked. The relative mean error and the estimated mean error for each model were calculated according to the following expressions, respectively:

$$P = \frac{100}{n} \sum \frac{\left|Y - \hat{Y}\right|}{Y} \quad (24)$$

$$SE = \sqrt{\frac{\sum \left(Y - \hat{Y}\right)^2}{GLR}} \quad (25)$$

in which,

Y - experimentally observed value

Y- value calculated by the model

n - number of experimental observations

GLR - degrees of freedom of the model (the number of observations minus the number of model parameters).

RESULTS AND DISCUSSION

The influence of air temperature on the drying kinetics of sunflower seeds can be seen in Figure 1. The initial water content of the sunflower seeds was 46.64 ± 1% (b.u.), and the final water content was 7.62; 7.76; 6.7 and 7.03% (b.u.) at 45, 55, 65 and 75, respectively. The drying time to reach the equilibrium water content for the sunflower seed sample was 7.5, 6.5, 5.5 and 4.0 hours at 45, 55 and 75 °C, respectively.

As expected, the drying period decreases the air temperature increasing, so there is a higher rate of water removal from the seed due to the increase in the humidity gradient between the seed and the air, decreasing the time needed to reduce the water content to the desired value, a factor observed by several researchers in other agricultural products: grapes (AZZOUZ et al., 2002); red pepper (AKPINAR et al., 2003); prickly pear (LAHSASNI et al., 2004); eggplant (ERTEKIN; YALDIZ, 2004); parboiled rice (MOHAPATRA; RAO, 2005); Adzuki beans (ALMEIDA et al., 2009); and jatropha (ULLMANN et al., 2010).

At each temperature interval of 10 °C, the drying time decreased by 13.33% (45 to 55 °C), 15.38% (55 to 65 °C) and 27.27% (65 to 75 °C), respectively. Raising the drying air temperature by 10 °C, from 65 to 75 °C, had the greatest effect on the drying time. In addition, the higher drying temperatures increased the difference in moisture between the periphery and the center of the seed, leading to excessive drying and deterioration of the seeds. The moisture ratio decreased steadily as drying progressed (Figure 1). It took 4.5, 2.7, 2.3 and 1.7 hours to obtain the final storage moisture of the seeds at drying temperatures of 45, 55, 65 and 75 °C, respectively. It was observed that it took half the total drying time to remove the residual moisture, due to slower diffusion. These observations are in line with previous results on drying agricultural products (RESENDE et al., 2008, ULLMANN et al., 2010; CORADI et al., 2014a, CORADI et al., 2014b). The constant rate period was not observed during the drying of sunflower seeds (Figure 1).

The drying process took place at a decreasing rate, except for a very short acceleration period at the beginning. With a higher water content, the increase in temperature had a considerable effect on drying rates, with a faster reduction in moisture at the start of drying than at the end. The rate of moisture migration from the inner surface to the outer surface decreases in the final stage of drying and, therefore, lower drying rates are observed (RAJKUMAR et al., 2007).

In practice, when high drying air temperatures are used, the quality of sunflower seeds decreases (ALMEIDA et al., 2009). Therefore, in order to optimize energy efficiency and obtain final seed quality, the use of drying temperatures below 65 °C is recommended.

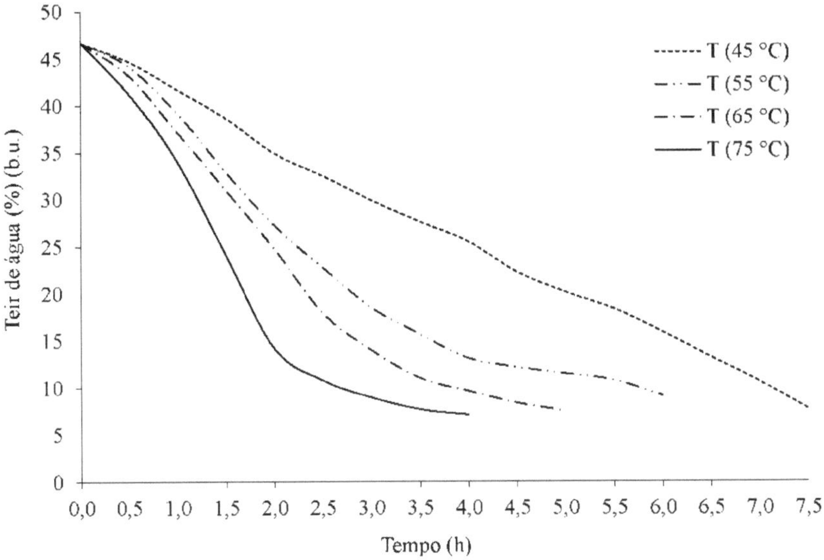

Figure 1. Sunflower seed drying curves.

The model coefficients are shown in Table 1. Regression analysis was used to establish the relationships between parameters and temperatures. Among the mathematical models evaluated, it was found that the coefficients of determination, average relative errors and average estimated errors (Table 3) determined that the Wang and Singh model presented the best fit to the experimental data to describe the drying process of sunflower seeds, in the temperature range of 45 to 75 °C.

Table 1. Parameters obtained from models fitted to sunflower seed drying results*.

Imaging models			
	T (^{0} C)	k	
	45	0,289430	
Exponential	55	0,464078	
	65	0,550907	
	75	0,759458	
	T (^{0} C)	k	n
	45	0,320876	0,921232

Page	55	0,483817	0,953479		
	65	0,524997	1,064947		
	75	0,729502 k	1,092830		
	T (^{0} C)		n		
	45	0,291157	0,921232		
Page	55	0,466978	0,953479		
Modified	65	0,546038	1,064947		
	75	0,749310	1,092830		
	T (^{0} C)	a	k		
	45	0,908492	0,259222		
Hendersone	55	0,918698	0,424069		
Pabis	65	0,928217	0,511096		
	75	0,918991	0,700129		
	T (^{0} C)	a	k	c	
	45	1,037938	0,189904	-0,15334	
Logarithmic	55	0,903337	0,450262	0,021476	
	65	0,978978	0,437512	-0,063986	
	75	0,947295	0,638264	-0,035473	
	T (^{0} C)	a	K0	b	k1
	45	0,45446	0,259220	0,454246	0,259220
Two Terms	55	0,459349	0,424060	0,459349	0,424069
	65	0,464108	0,511096	0,464108	0,511096
	75	0,459495	0,700129	0,459495	0,700129
	T (^{0} C)	a	k		
Exponential	45	0,520959	0,400013		
Two Terms	55	0,498911	0,670505		

	65	1,512833	0,659597				
	75	1,586729	0,946706				
	T (^{0} C)	a	b				
	45	-0,238528	0,016447				
Wang and Sing	55	-0,374451	0,038879				
	65	-0,432253	0,050702				
	75	-0,581200	0,089508				
	T (^{0} C)	a	k	b	k0	c	k1
Henderson e Modified Pabis	45	0,302837	0,259220	0,302837	0,259220	0,302837	0,259220
	55	0,306233	0,424060	0,306233	0,424060	0,306233	0,424060
	65	0,309406	0,511096	0,309406	0,511096	0,309406	0,511096
	75	0,306330	0,700129	0,306330	0,700129	0,306330	0,700129
	T (^{0} C)	a	k	n	b		
	45	0,884174	0,211266	1,013423	-0,01038		
Midilli	55	0,882623	0,354036	1,385108	0,017457		
	65	0,868485	0,374479	1,491149	0,013655		
	75	0,867934	0,592848	1,572112	0,019016		
	T (^{0} C)	a	k	b			
	45	0,064322	11,54381	0,023224			
Getting closer by Diffusion	55	0,998079	0,488154	-1,129390			
	65	-2,421390	0,817318	0,883372			
	75	9,352686	1,084651	1,049122			

*All estimated coefficients were significant at 5% probability using the t-test.

Table 2. Coefficient of determination (R^2), relative mean error (P), estimated mean error (SE) values of sunflower seed drying and function of different temperatures.

	45 C^0	55 C^0	65 C^0	75 C^0

		R^2 (%)		
Exponential	98,66	98,73	98,66	98,37
Page	98,80	98,77	98,73	98,47
Page Modified	98,80	98,77	98,73	98,47
Henderson and Pabis	99,67	99,36	99,15	98,98
Logarithmic	99,91	99,38	99,28	99,04
Two Terms	99,67	99,36	99,15	98,98
Two-Term Exponential	98,70	98,83	98,73	98,48
Wang and Singh	97,80	98,51	98,75	98,55
Henderson and Pabis Modified	99,67	99,36	99,15	98,98
Midilli	99,61	99,94	99,88	99,90
Diffusion Approximation	98,93	98,98	98,73	98,46
		IF (decimal)		
Exponential	0,0491	0,0704	0,0764	0,0786
Page	0,0629	0,0804	0,0706	0,0702
Page Modified	0,0629	0,0804	0,0706	0,0702
Henderson and Pabis	0,0761	0,0936	0,0978	0,1029
Logarithmic	0,0691	0,0955	0,0912	0,0985
Two Terms	0,0822	0,1035	0,1109	0,1218
Two-Term Exponential	0,0586	0,0822	0,0696	0,0685
Wang and Singh	0,0487	0,0745	0,0634	0,0661
Henderson and Pabis Modified	0,0900	0,1174	0,1312	0,1573
Midilli	0,0745	0,0969	0,0925	0,0991
Diffusion Approximation	0,0676	0,0871	0,0740	0,0747
	P (%)			
Exponential	14,24	7,67	5,84	8,51
Page	12,27	8,20	5,23	7,55

Page Modified	12,27	8,20	5,23	7,55
Henderson and Pabis	0,62	6,14	3,25	5,08
Logarithmic	4,15	6,33	2,69	4,75
Two Terms	0,62	6,14	3,25	5,08
Exponential Two Terms	13,72	8,40	5,20	7,74
Wang and Singh	23,56	10,15	8,76	10,94
Henderson and Pabis Modified	7,74	6,14	8,25	5,08
Midilli	4,18	6,95	8,79	4,70
Diffusion Approximation	10,45	8,11	5,88	7,10
		Waste distribution		
Exponential	A	A	A	A
Page	A	A	A	A
Page Modified	A	A	A	A
Henderson and Pabis	A	A	A	A
Logarithmic	A	A	A	A
Two Terms	A	A	A	A
Two-Term Exponential	A	A	A	A
Wang and Singh	A	A	A	A
Henderson and Pabis Modified	A	A	A	A
Midilli	A	A	A	A
Diffusion Approximation	A	A	A	A

Figure 2 shows the experimental data for water content versus drying time for each drying air temperature adjusted to the Wang and Singh model. Figure 3 shows the relationship between the experimental and estimated values and the satisfactory fit of the Wang and Singh model to describe the drying of sunflower seeds. The responses shown in Figure 5 demonstrate that the data points follow a straight line at an angle of 45°, meaning that the model is suitable for describing the thin layer of sunflower seed drying. A similar approach to selecting a drying model for biological products was observed by RESENDE et al. (2008), GONELI et al. (2011) and CORADI et al. (2014a).

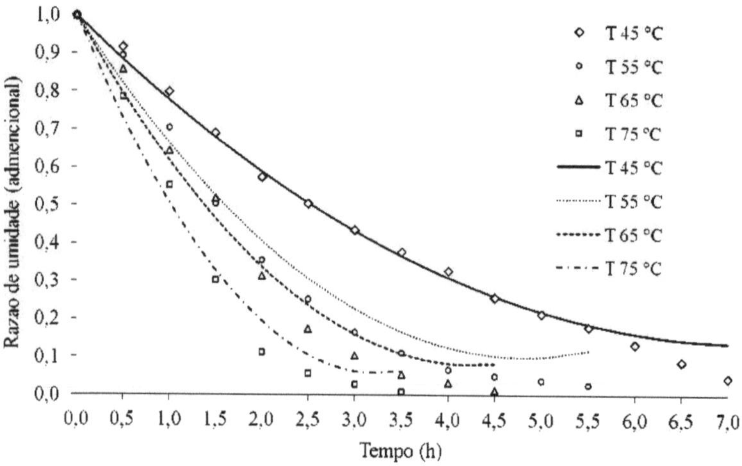

Figure 2 - Sunflower seed moisture ratio fitted to the Wang and Singh model.

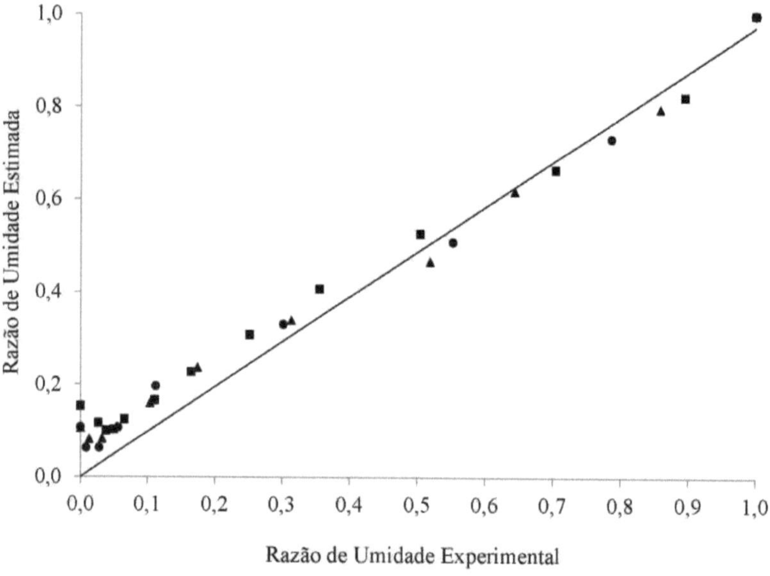

Figura 3. Experimental data and estimate of the moisture ratio, calculated using the Wang and Singh model.

The average values of the effective diffusivities of the sunflower seeds in the drying process from 45 to 75 °C ranged from 2.83-2.93 x 10^{-6} m^2 s^{-1} (Figure 4). It can also be seen in Figure 4 that the effective diffusivity increased with increasing temperature. The results are in line with those found by WANG et al. (2007); RESENDE et al. (2008); GONELI et al. (2011); CORADI et al. (2014b).

The effective diffusivities evaluated were within the range of published values, varying from 10^{-11} to 10^{-6} for food materials (WANG et al., 2006; RESENDE et al., 2008, CORADI et al., 2014a). Figure 5 shows the D values presented as "ln D" described as a function of the reciprocal absolute temperature (1/T). The straight line obtained indicates the uniformity of variation of diffusivity with temperature.

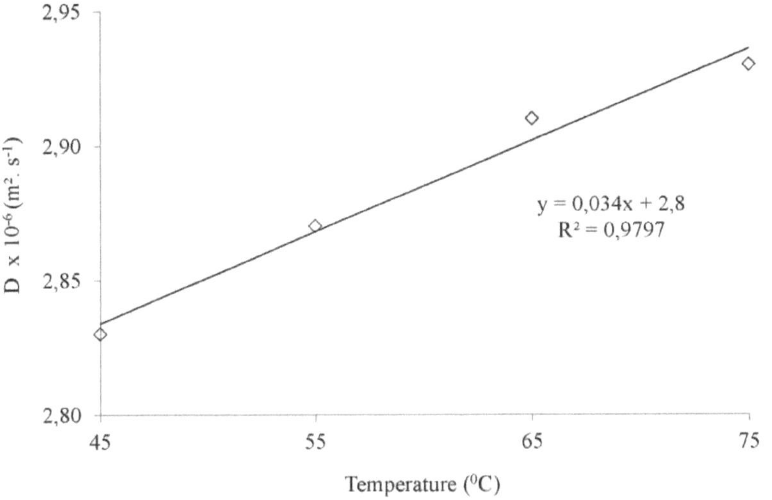

Figura 4. Average values of the effective diffusion coefficient (m2 s^{-1}) due to different air temperatures when drying sunflower seeds.

The latent heat of vaporization curves for sunflower seeds at 45, 55, 65 and 75 °C are shown in Figure 6. The latent heat of vaporization ranged from 3,725.176 to 3,773.082 kJ kg^{-1} , while as the equilibrium water content decreased, there was an increase in the energy required to evaporate the water from the corn kernels. According to BROOKER et al. (1992), water content and temperature are the main factors influencing the value of the latent vapor of water vaporization of the product. It was found that with an increase in temperature for the same water content, there is a reduction in the latent heat of vaporization. ALMEIDA et al. (2000) studied the latent heat of vaporization of cotton seeds and lint and observed the same behavior.

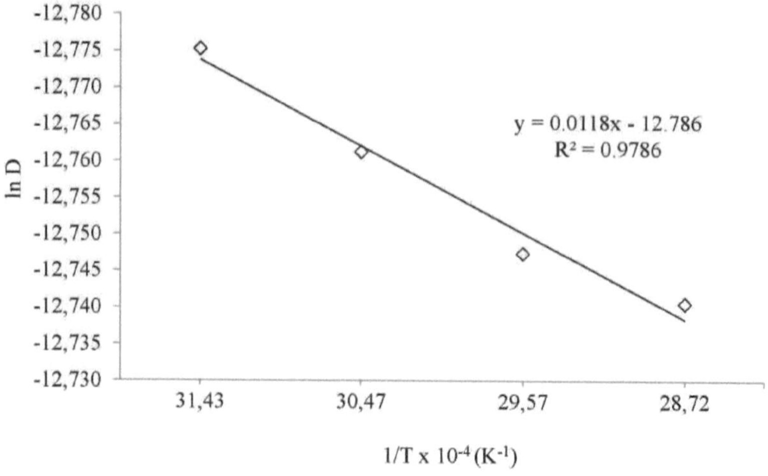

Figura 5. Representation of the Arrhenius relation for effective diffusivity and air drying of sunflower seeds.

The same results were observed in CORADI et al. (2014a) and CORADI et al. (2014b), who studied the latent heat of vaporization in coffee beans and lemongrass plants.

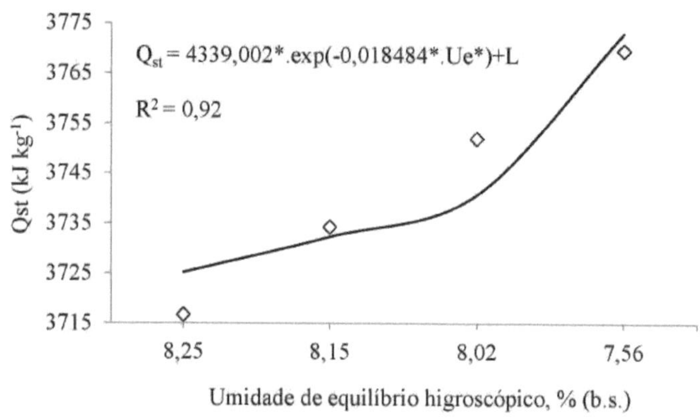

Figura 6. Experimental values and estimated integral isosteric heat of sorption as a function of equilibrium moisture content.

The Correa et al. model (Table 3) was the most suitable for the mathematical description of the sunflower seed shrinkage phenomenon (Figure 7), showing a high coefficient of determination and low values for the average and estimated relative errors, while all the models showed a good distribution and regular residuals (Tables 3 and 4). Figure 7 shows that when the water content was reduced from 0.86 to 0.09 (kg kg^{-1} of dry matter) there was a marked reduction in seed volume, relative to the initial volume, at temperatures of 75, 65, 55 and 45 °C (Figure 7).

Table 3. Estimated parameters, coefficient of determination (R^2) and distribution of residuals of the mathematical models used to describe the shrinkage of sunflower seeds at different drying air temperatures.

Mathematical models	Parameter estimation	R2	Waste distribution
Temperature 45 °C			
Bala and Woods	a= 0,92912 b= -18,0258	68,23	T
Rahman	a=0,75374 b= 0,52211	98,35	A
Correa et al. (2002)	a= 1,77125 b= -0,4918	98,86	A
Exponential	a= 0,79186 b= 0,58681	98,77	A
Temperature 55 °C			
Bala and Woods	a= 0,90889 b= -25,7737	28,94	T
Rahman	a=0,78178 b= 0,419414	87,10	A
Correa et al. (2002)	a= 1,6682 b= -0,41392	90,75	A
Exponential	a= 0,78537 b= 0,478184	88,09	A
Temperature 65 °C			
Bala and Woods	a= 0,91238 b= -18,7278	62,48	T

Rahman	a=0,74456 b= 0,506803	96,00	A
Correa et al. (2002)	a= 1.79890 b=-0.4999	98,03	A
Exponential	a= 0,75167 b= 0,581219	96,65	A
Temperature 75 °C			
Bala and Woods	a= 0.90963 b=-17.8313	66,13	T
Rahman	a=0,73223 b= 0,528912	96,29	A
Correa et al. (2002)	a= 1,84398 b= -0,52740	98,13	A
Exponential	a= 0,74023 b= 0,610357	96,90	A

Table 4 - Average relative error values (P), estimated average error values (EP) of sunflower seeds dried at different temperatures

	45 C^0	55 C^0	65 C^0	75 C^0
		IF (decimal)		
Bala and Woods	0,051336	0,060971	0,0545230	0,054501
Rahman	0,012667	0,031287	0,0195408	0,019595
Correa et al. (2002)	0,007480	0,026750	0,0137760	0,013976
Exponential	0,010959	0,030143	0,0179071	0,017941
		P (%)		
Bala and Woods	5,84627	1,926450	5,1696970	5,595600
Rahman	0,75000	0,130000	0,1010000	0,760000
Correa et al. (2002)	0,32463	0,653270	0,4535350	0,414200
Exponential	2,20796	4,005950	3,3492420	3,357070

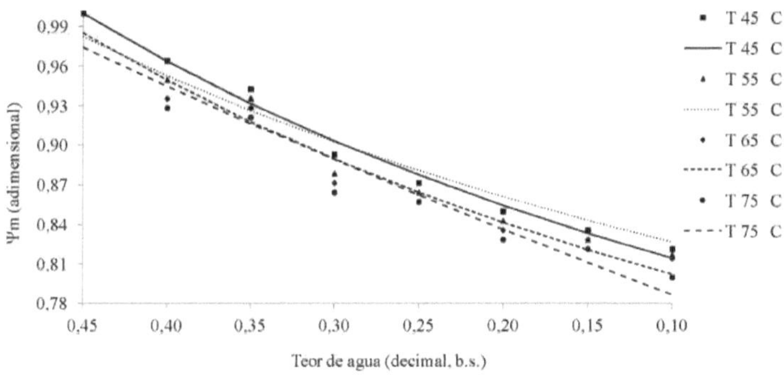

Figura 7. Volumetric contraction of sunflower seeds adjusted to the Corrêa et al. model due to drying at different air temperatures.

The drying temperature of 45 °C reduced the volume of the seeds by the smallest percentage, due to the low rate of water removal, causing the integument to stiffen, which made it difficult for the seeds to contract. Temperatures between 65 and 75 °C had the greatest influence on seed volume, since the water was removed quickly and the seeds contracted their orthogonal axes as the water content decreased. SIQUEIRA et al. (2011) found that the decrease in the geometric mean diameter of the seeds is proportional to the reduction in water content and also depends on the drying condition, i.e. the air drying temperature.

CONCLUSIONS

The drying time decreased significantly as the temperature of the drying medium increased. The Wang and Sing model was the best fit to the experimental drying data. The effective diffusivity of sunflower seeds ranged from 2.83 to 2.93x10^{-6} m^2 s^{-1} , while the isosteric heat of adsorption ranged from 3,725.176 to 3,773.082 kJ kg^{-1} . Reducing the water content influenced the volumetric contraction of sunflower seeds, especially at temperatures above 55 °C. The Correa et al. model was the best fit to the experimental data on sunflower seed volume contraction.

BIBLIOGRAPHICAL REFERENCES

ALMEIDA, F. de A. C.; MATA, M. E. R. M. C.; CORRÊA, P. C. Adsorption isotherms and latent heat of vaporization of cotton seeds, with and without linter. **Revista Oleaginosas e Fibrosas**, Campina Grande, v. 4, n. 3, p. 225-234, 2000.

ALMEIDA, D. P. RESENDE, O.; COSTA, L.M.; MENDES, U. C.; SALES, J. F. Drying kinetics of adzuki beans (*Vigna angularis*). **Global Science and Technology,** v. 2, n. 1, p. 7283, 2009.

BRAZIL. Ministry of Agriculture and Agrarian Reform. **Rules for seed analysis.** Brasilia: SNDA/DNDV/CLAV, 2009.

BROOKER, D. B.; BAKKER-ARKEMA, F. W.; HALL, C. W. **Drying and storage of grains and oilseeds.** Westport: The AVI Publishing Company, 1992, 450p.

CANEPPELE, F. L.; SERAPHIM, O. J. Application of fuzzy theory in the control of alternative energy generation systems. **Energia na Agricultura**, v. 25, n. 3, p. 24-41, 2010.

CANEPPELE, F. E; SERAPHIM, o. J. Analysis of energy efficiency in wood industries using fuzzy logic. **Energia na Agricultura**, v. 28, n. 2, p. 95-102, 2013.

CORADI, P. C.; MELO, E. C.; ROCHA, R. P. Evaluation of electrical conductivity as a quality parameter of lemongrass leaves (*Cymbopogon Citratus* Stapf) submitted to drying process. **Drying Technology (Online),** v. 32, n. 8, p. 969-980, 2014a.

CORADI, P. C.; BORÉM, F. M.; REINATO, C. H. Coffee cherries drying process and the influence of environment relative humidity in the mathematical modeling, moisture content, and enthalpy of vaporization. **Energia na Agricultura (UNESP. Botucatù. CD-Rom)**, v. 29, n. 2, p. 148-157, 2014b.

CORRÊA, P. C.; AFONSO JR., P. C.; MARTINS, P. M.; MELO, E. C.; RADUNZ, L. L. Mathematical model for representing the hygroscopicity of medicinal plants. **Revista Brasileira de Armazenamento**, v. 27, n. 1, p. 9-13, 2002.

GONELI, A. L. D. CORRÊA, P. C.; MAGALHÂES, F. E. A.; BAPTESTINI, F. M. Volumetric contraction and shape of castor bean fruits during drying. **Acta Scientiarum. Agronomy,** v. 33, n.1, p. 1-8, 2011.

ERTEKIN C, YALDIZ, O. Drying of eggplant and selection of a suitable thin layer drying model. **J. Food Eng.** v. 63, n. 3, p. 349-359, 2004.

LAHSASNI, S. KOUHILA, M.; MAHROUZ, M.; JAOUHARI, J. T. Drying kinetics of prickly pear fruit (*Opuntia ficus indica*). **Journal of Food Engineering**. v. 2, n. 2, p. 73-179, 2004.

MAYOR, L.; SERENO, A. M. Modeling shrinkage during convective drying of food materials: a review. **Journal of Food Engineering,** v. 61, n. 3, p. 373-386, 2004.

MIR, S. A.; BOSCO, S. J. D.; SUNOOJ, K. V. Evaluation of physical properties of rice cultivars grown in the temperate region of India. **International Food Research Journal,** v. 20, n. 4, p. 1521-1527, 2013.

MOHAPATRA, D.; RAO, P. S. A thin layer drying model of parboiled wheat. **Journal of Food Engineering.** v. 34, n. 4, p. 513-518, 2005.

MOSHENIN, N. N. **Physical properties of plant and animal materials.** New York, USA. 1986, 841p.

OLIVA, A. C. E.; BIAGGIONI, M. A. M.; CAVARIANI, C. Immediate effect of drying method on crambe seed quality. **Energia Agricultura**, v. 27, n. 3, p. 16-30, 2012. PESAGRO. **Agricultural Research Company of Rio de Janeiro**, available at,

http://www.pesagro.rj.gov.br/downloads/maisnoticia_jul07.pdf, consulted on 15/07/2007.

RAJKUMAR, P.; KAILAPPAN, P.; VISWANATHAN, R.; RAGHAVAN, G. S. V. Drying characteristics of foamed alphonso mango pulp in a continuous type foam mat dryer. **J. Food Eng.** v. 79, n. 4, p. 1452-1459, 2007.

RESENDE, O. CORRÊA, P. C.; GONELI, A. L. D.; BOTELHO, F. M.; RODRIGUES, S.Mathematical modeling of the drying process of two varieties of beans (*Phaseolus vulgaris* L.). **Revista Brasileira de Produtos Agroindustriais,** v. 10, n. 1, p. 17-26, 2008. SILVA, J. S. **Secagem e Armazenagem de Produtos** Agricolas. Viçosa, Aprenda Facil, 2008. SILVA, M. A. P.; BIOAGGIONI, M. A. M.; SPEROTTO, F. C. S.; BEZERRA, P. H. S.; BRANDÂO, F. J. B. Quality of crude oil from crambe grains (Crambe abyssinica Hochst) under different drying methods). **Energia na Agricultura**. v. 28, n. 3, p. 193-199, 2013. SIQUEIRA, V. C.; RESENDE, O.; CHAVES, T. H. Volumetric contraction of pine nuts during drying at different temperatures. **Revista Brasileira de Armazenamento**, Viçosa, v. 36, n. 2, p. 171-178, 2011.

SIQUEIRA, J. A. C.; GABRIEL FILHO, L. R. A.; CANEPPELE, F. L.; SERAPHIM, O. J. Characteristics of hybrid systems for electricity generation and applications in wind-photovoltaic systems. **Brazilian Journal of Biosystems Engineering**, v. 8, n. 1, p. 84-96, 2014.

ULLMANN, R. RESENDE, O.; SALES, J. F.; CHAVES, T. H. Quality of pine seeds subjected to drying in different air conditions. **Revista Ciência Agronômica**. v. 41, n. 3, p. 442-447. 2010.

YALÇIN, Y.; OZARSLAN, C.; AKBAb, T. Physical properties of pea (*Pisum sativum*) seed.

Journal of Food Engineering, v. 79, n. 2, p. 731-735, 2007.

WANG, Z.; SUN, J.; LIAO, X.; CHEN, F.; ZHAO, G.; WU, J.; HU, X. Mathematical modeling on hot air drying of thin layer apple pomace. **Food Res. Int.** v. 40, n. 1, p. 39-46, 2006.

WANG, Z.; SUN, J.; LIAO, X.; CHEN, F.; ZHAO, G.; WU, J.; HU, X. Mathematical modeling on thin layer microwave drying of apple pomace with and without hot air predrying. **J. Food Eng.** v. 80, n. 2, p. 536-544, 2007.

CHAPTER II

EFFECTS OF DRYING AND STORAGE CONDITIONS ON THE QUALITY OF SUNFLOWER SEEDS

Abstract: The viability of seeds depends on suitable harvesting and post-harvest conditions, such as storage. The aim of this study was to evaluate the physiological potential of sunflower seeds under different storage conditions (25^0 C and 50%, 20^0 C and 60%, 30^0 C and 40% RH) using different packaging (permeable and impermeable). The physiological evaluations of the seeds were carried out at zero, three and six months of storage and analyzed using the mean test, compared by the Tukey test at 5% probability. The best quality results were obtained for the impermeable packaging under conditions of 20 °C and 60% RH and 25 °C and 50% RH. The physiological potential of sunflower seeds was affected by drying air temperature and storage conditions. Drying air temperature above 45 °C negatively affected the quality of sunflower seeds. Permeable and impermeable packaging did not guarantee the quality of sunflower seeds throughout storage. The quality of sunflower seeds stored for longer than three months was negatively affected. The conditions of 20 °C and 60% RH were the most favorable for preserving the physiological quality of sunflower seeds over the storage period.

Keywords: Preservation, *Helianthus annus* L., quality.

INTRODUCTION

The sunflower (*Helianthus annus* L.), a plant of the Asteraceae family, is native to North America and, until the 17th century, was cultivated as an ornamental and medicinal plant. In the 18th century, it was selected in Russia as an oil production plant, but only gained economic importance after the Second World War. It is currently the world's second largest source of edible vegetable oil. Sunflower can be an alternative supply of raw materials for oil extraction, even in times when the oil extraction industry is idle. There is also the possibility of using its oil as fuel (BENNETT, 2011). Thus, most seeds after harvest have unsuitable characteristics for storage, mainly due to the high water content that needs to be dry for storage (BAILEY, 1992; MARTINS et al., 2009).

The water content recommended for safe storage is a very important factor and depends on the species, the environmental conditions of the region, the period and the type of packaging used. Some physical factors (temperature, humidity and mechanical damage) and biological factors (insects, dust mites and microorganisms) affect the preservation of stored grain. The water content greatly influences the quality of the stored product, and seeds with a high water content favor the growth of microorganisms, insects and mites (PRADHAN et al., 2009). Storage can reduce the longevity of seeds mainly due to the effects of external factors such as temperature and relative humidity, altering the water content and the speed of biochemical processes in seeds. However, when seeds are stored at low temperatures, there is less chance of deterioration.

The most obvious physiological changes in seeds due to the effects of drying and storage are observed during germination and initial seedling development. The disintegration of the seed membrane system during storage has been an observed factor in terms of quality, altering the chemical instability of lipids and causing a decrease in seed performance (CLEMENT et al., 2011; SANTOS et al., 2012).

It is therefore essential to evaluate seed germination, speed of emergence and seedling development throughout seed storage. The aim of this study was to evaluate the physiological potential of sunflower seeds under different storage conditions (25 °C and 50%, 20 °C and 60%, 30 °C and 40% RH), using different packaging (permeable and impermeable) for six months.

MATERIALS AND METHODS

The experimental work was carried out at the Federal University of Mato Grosso do Sul (UFMS), Chapadào do Sul Campus (CPCS), located in the municipality of Chapadào do Sul (MS). The sunflower seeds were harvested at random, and the possession of specimens of manual separation of impurities was carried out with damaged seeds. The seeds were then dried in a convection oven with forced ventilation at temperatures of 45, 55, 65 and 75 °C. The seeds were dried until they reached the water content of 10.5% (b.u.) recommended for storage. Three tests were carried out for each drying air temperature and 5 kg of sunflower seeds were used for each test. Temperature and relative humidity were monitored throughout the drying process using a psychrometer.

After drying, the seeds were stored in three different environments: a natural environment (25°C and 50% RH) and two climate chambers (20°C and 60% RH, 30°C and 40% RH), two types of packaging (permeable and impermeable) for a period of six months.

Over time, the stored seeds were monitored for temperature and relative humidity using a psychrometer. To assess the quality of the sunflower seeds after drying, germination, accelerated ageing, speed of emergence and seedling growth tests were carried out.

The germination test was carried out with four replicates of 30 seeds from each treatment, on "Germitest" type paper rolls in a "Mangesdorf" type germinator, at a constant temperature of 25 ± 2 °C. The amount of water added was equivalent to 2.5 times the mass of dry substrate, in order to ensure adequate wetting and, consequently, standardization of the test. Interpretations were made from the 4th day after sowing until the 10th day according to the criteria established in the Rules for Seed Analysis (BRASIL, 2009).

The accelerated ageing test was carried out using the seedling method, in a BOD plastic box chamber, where four grams of seeds were placed on a screen attached to an aluminum gerbox, containing 40 mL of distilled water at the bottom and then taken to the ageing chamber (BOD) set at 41 °C for 48, 72 and 96 hours. The seeds were then submitted to the germination test and evaluated on the sixth day after installation and the results expressed as a percentage of normal seedlings (BRASIL, 2009).

The emergence test was carried out by sowing sunflower seeds in substrate trays, with four replicates of 50 seeds for each sample. The emerged seedlings were counted every day for the first 5 days (MAGUIRE, 1962). The length of the seedling roots and the sunflower seeds were evaluated in five samples of 20 seeds from each treatment, distributed on paper towel rolls moistened with distilled water in a ratio of 2.5 to 1 (mL of distilled water to dry mass of paper) and kept in a germination chamber at 25 °C for five days (NAKAGAWA, 1999). A line was drawn longitudinally across the upper third of the moistened paper towel. The length of the primary root and the seedling considered normal (BRASIL, 2009) were determined at the end of the fifth day using a millimeter ruler.

The experimental design was entirely randomized (3x3x2), with three drying air temperatures, three storage times and two types of packaging, with three replications per treatment. Analysis of variance was carried out and the means were compared using the Tukey test at 5% probability.

Results and discussion

Physiological evaluation is related to the seed's ability to perform its vital functions, characterized by vigour, germination and longevity. A reduction in quality is generally reflected in a decrease in the percentage of germination, abnormal seedlings and a reduction in seedling vigor (TOLEDO et al., 2009). Tables 1, 2 and 3 show the results of sunflower seed germination.

Table 1. Sunflower seed germination (%) according to drying air temperature and storage conditions, for each type of packaging and storage time.

Storage time (months)	Packaging	Conditions storage	Drying air temperature 45 °C	55 °C	65 °C	75 °C
0	-	-	100 a	94 b	78 d	84 c
		20 °C / 60%	96 Aa	82 Bb	82 Cb	86 Bb
	Waterproof	25 °C / 50%	94 Ab	100 Aa	88 Bc	10 Ca
3		30 °C / 40%	96 Ab	100 Aa	96 Ab	92 Ac

Storage time (months)	Packaging	Storage conditions	45 °C	55 °C	65 °C	75 °C
	Permeable	20 °C / 60%	94 Ab	100 Aa	96 Ab	96 Ab
		25 °C / 50%	96 Aa	96 Ba	96 Aa	96 Aa
		30 °C / 40%	0 Ba	0 Ca	0 Ba	0 Ba
6	Waterproof	20 °C / 60%	19 Bc	20 Ac	41 Aa	31 Bc
		25 °C / 50%	22 Ab	17 Bc	30 Ca	28 Ca
		30 °C / 40%	14 Cc	0 Cd	37 Bb	45 Aa
	Pemieable	20 °C / 60%	12 Bb	8 Bc	19 Ba	21 Aa
		25 °C / 50%	32 Aa	32 Aa	26 Ab	0 Ba
		30 °C / 40%	0 Ca	0 Ca	0 Ca	0 Ba

Averages followed by a capital letter in the column for each storage condition and a small letter in the rows for each drying air temperature do not differ at 5% probability. Coefficient of variation (CV) = 5.23%.

Table 2. Sunflower seed germination (%) according to drying air temperature and storage time, for each type of packaging and storage conditions.

Storage time (months)	Packaging	Storage conditions	Drying air temperature			
			45 °C	55 °C	65 °C	75 °C
0		-	100 Aa	94 Ab	78 Bd	84 Ac
3	Waterproof	20 °C / 60%	96 Ba	82 Bc	82 Ac	86 Ab
6		20 °C / 60%	19 Cc	20 Cc	41 Ca	31 Bb
0		-	100 Aa	94 Bb	78 Bd	84 Bc
3	Permeable	20 °C / 60%	94 Bb	100 Aa	96 Ab	96 Ab
6		20 °C / 60%	12 Cb	8Cc	19 Ca	21 Ca
0		-	100 Aa	94 Bb	78 Bd	84 Bc
3	Waterproof	25 °C / 50%	94 Bb	100 Aa	88 Ac	100 Aa
6		25 °C / 50%	22 Cc	17Cd	30 Ca	28 Cb
0		-	100 Aa	94 Ab	78 Bd	84 Bc

3	Permeable	25 °C / 50%	96 Ba	96 Aa	96 Aa	96 Aa
6		25 °C / 50%	32 Ca	32 Ba	26 Cb	0 Cc
0		-	100 Aa	94 Bb	78 Bd	84 Bc
3	Waterproof	30 °C / 40%	96 Bb	100 Aa	96 Ab	92 Ac
6		30 °C / 40%	14 Cc	0Cd	37 Cb	45 Ca
0		-	100 Aa	94 Ab	78 Ad	84 Ac
3	Permeable	30 °C / 40%	0Ba	0 Ba	0 Ba	0 Ba
6		30 °C / 40%	0Ba	0 Ba	0 Ba	0 Ba

Means followed by a capital letter in the column for each storage condition and a small letter in the rows for each drying air temperature do not differ at 5% probability. Coefficient of variation (CV) = 4.41%.

It was observed that drying temperatures above 45 °C negatively affected seed germination. Meanwhile, the effects of storage conditions were greater over time for seeds dried at 55, 65 and 75 °C.

Table 3. Sunflower seed germination (%) versus storage time and type of packaging for each of the temperature and drying air storage conditions.

Drying air temperature	Storage conditions	Packaging	Storage time (months)		
			0	3	6
45 °C	20 °C / 60%	Waterproof	100 Aa	96 Ab	19 Ac
		Permeable	100 Aa	94 Ab	12 Bc
	25 °C / 50%	Waterproof	100 Aa	94 Ab	22 Bc
		Permeable	100 Aa	96 Ab	32 Ac
	30 °C / 40%	Waterproof	100 Aa	96 Ab	14 Ac
		Permeable	100 Aa	0 Bb	0 Bb
55 °C	20 °C / 60%	Waterproof	94 Aa	82 Bb	20 Ac
		Permeable	94 Ab	100 Aa	8 Bc
		Waterproof	94 Ab	100 Aa	17 Bc

	25 °C / 50%	Permeable	94 Aa	96 Ba	32 Ac
		Waterproof	94 Ab	100 Aa	0 Ac
	30 °C / 40%	Permeable	94 Aa	0Bb	0 Ac
		Waterproof	78 Ab	82 Ba	41 Ac
	20 °C / 60%	Permeable	78 Ab	96 Aa	19 Bc
65 °C		Waterproof	78 Ab	88 Ba	30 Ac
	25 °C / 50%	Permeable	78 Ab	96 Aa	26 Bc
		Waterproof	78 Ab	96 Aa	37 Ac
	30 °C / 40%	Permeable	78 Aa	0 Bb	0 Bb
		Waterproof	84 Aa	86 Ba	31 Ab
	20 °C / 60%	Permeable	84 Ab	96 Aa	21 Bc
75 °C		Waterproof	84 Ab	100 Aa	28 Ac
	25 °C / 50%	Permeable	84 Ab	96 Ba	0 Bc
		Waterproof	84 Ab	92 Aa	45 Ac
	30 °C / 40%	Permeable	84 Aa	0 Bb	0 Bb

Means followed by a capital letter in the column for each storage condition and a small letter in the rows for each drying air temperature do not differ at 5% probability. Coefficient of variation (CV) = 6.44%.

In the first three months, it was found that the storage conditions of 25 °C and 50% RH in air provided the highest percentage of germinated seeds. However, after the third month of storage, the conditions of 20 °C and 60% RH showed the best results for seed germination (Table 1, 2 and 3). Almeida et al. (2010) observed a linear decrease in the germination of five oilseed species (cotton, peanuts, soybeans, sunflower seeds and castor beans) as the storage period increased under ambient conditions, data which agrees with that obtained by Fessel et al. (2010), who stated that the physiological quality (germination and vigor) of rice, corn and bean seeds decreased over the storage period.

According to Carvalho and Nakagawa (2000), seeds show greater germination and vigor during physiological maturation. From this moment on, gradual physiological and biochemical changes inevitably occur, causing deterioration and a gradual loss of vigor. According to Tonin and Perez

(2006), seed quality is an extremely important factor in achieving productivity. The packaging used in storage should help slow down the deterioration process by maintaining the initial water content of the stored seeds in order to reduce respiration. Thus, a 100% reduction in the germination of seeds stored in permeable packaging conditions of 30°C and 40% RH of air was observed.

There was also a significant difference (p<0.05) in germination between seeds stored in permeable and impermeable packaging and which were dried at temperatures above 45 °C. The percentage of germinated seeds was higher in impermeable packaging. According to Marcos Filho (2005) the deterioration of seeds is associated with the characteristics of their containers, depending on the greater or lesser ease of water vapor exchange between the seeds and the atmosphere and the environmental conditions in which they are stored. Similar results were observed with the germination and accelerated ageing of sunflower seeds (Table 4). The drying temperatures of 45 and 65 °C affected the quality of the seeds the least, and the storage conditions of 20 °C and 60%, 25 °C and 50% were the best for preserving the seeds, regardless of the type of packaging (Table 5, 6 and 7).

There were significant variations (P<0.05) in the quality of stored sunflower seeds, in permeable and impermeable packaging (Table 6). The permeable packages showed higher percentages of germinated seeds. Seeds stored in permeable packaging react with the humid environment, encouraging the exchange of water vapor between the seeds and the atmosphere, increasing the germination percentage. According to Marcos Filho (2005), seed deterioration is associated with the characteristics of the respective containers, depending on the water vapor exchange conditions between the atmosphere and the seeds and the conditions of the environment in which they are stored.

Table 4: Accelerated ageing of sunflower seeds (%) versus air drying temperature and storage conditions for each type of packaging and storage time.

Storage time (months)	Packaging	Storage conditions	Drying air temperature			
			45 °C	55 °C	65 °C	75 °C
0	-	-	92 b	96 a	92 b	98 a
		20 °C / 60%	90 Aa	82 Bb	52 Bd	70 Bc
	Waterproof	25 °C / 50%	76 Ba	64 Cc	70 Ab	60 Cc
3		30 °C / 40%	78 Bc	96 Aa	72 Ad	82 Ab
		20 °C / 60%	86 Ba	86 Aa	68 Bb	68 Bb

	Permeable	25 °C / 50%	96 Aa	82 Bc	92 Ab	90 Ab
		30 °C / 40%	0 Ca	0 Ca	0 Ca	0 Ca
		20 °C / 60%	34 Bb	32 Bb	37 Aa	36 Ba
	Waterproof	25 °C / 50%	39 Aa	37 Aa	36 Ab	34 Bb
6		30 °C / 40%	7 Cb	0 Cc	35 Aa	42 Aa
		20 °C / 60%	36 Aa	31 Ab	30 Ab	29 Ab
	Permeable	25 °C / 50%	32 Ba	31 Aa	33 Aa	0 Bb
		30 °C / 40%	0 Ca	0 Ba	0 Ba	0 Ba

Averages followed by a capital letter in the column for each storage condition and a small letter in the rows for each drying air temperature do not differ at 5% probability. Coefficient of variation (CV) = 5.36%.

Time (Table 5) and storage conditions were the main variables that affected the analysis of seed quality in the accelerated ageing of seeds, with significant results for all treatments. In permeable packages stored at 30 °C and 40% RH, there was a significant reduction in seed germination after three months of storage. Storage becomes a critical stage in maintaining seed quality until sowing or marketing (AZEVEDO et al., 2003). The packaging used for storing seeds should help slow down the deterioration process by maintaining the initial water content of the stored seeds in order to reduce respiration (TONIN and PEREZ, 2006).

Table 5: Accelerated ageing of sunflower seeds (%) according to drying air temperature and storage time for each type of packaging and storage conditions.

Storage time (months)	Packaging	Storage conditions	Drying air temperature			
			45 °C	55 °C	65 °C	75 °C
0		-	92 Ab	96 Aa	92 Ab	98 Aa
3	Waterproof	20 °C / 60%	90 Aa	82 Ab	52 Bd	70 Bc
6		20 °C / 60%	34 Bb	32 Bb	37 Ca	36 Ca
0		-	92 Ab	96 Aa	92 Ab	98 Aa
3	Permeable	20 °C / 60%	86 Ba	86 Ba	68 Bb	68 Bb
6		20 °C / 60%	36 Ca	31 Cb	30 Cb	29 Cb

0		-	92 Ab	96 Aa	92 Ab	98 Aa
3	Waterproof	25 °C / 50%	76 Ba	64 Bc	70 Bb	60 Bc
6		25 °C / 50%	39 Ca	37 Cb	36 Cb	34 Cb
0		-	92 Bb	96 Aa	92 Ab	98 Aa
3	Permeable	25 °C / 50%	96 Aa	82 Bc	92 Ab	90 Bc
6		25 °C / 50%	32 Ca	31 Ca	33 Ca	0 Cb
0		-	92 Ab	96 Aa	92 Ab	98 Aa
3	Waterproof	30 °C / 40%	78 Bc	96 Aa	72 Bd	82 Bb
6		30 °C / 40%	7 Cc	0 Bd	35 Cb	42 Ca
0		-	92 Ac	96 Ab	92 Ac	98 Aa
3	Permeable	30 °C / 40%	0 Ba	0 Ba	0 Ba	0 Ba
6		30 °C / 40%	0 Ba	0 Ba	0 Ba	0 Ba

Averages followed by a capital letter in the column for each storage condition and a small letter in the rows for each drying air temperature do not differ at 5% probability. Coefficient of variation (CV) = 4.91%.

As for the seed vigor index determined by the speed of emergence (Figure 7), there was a significant difference between the tests, which shows that the emergence rate was affected by the respective treatments when subjected to storage periods. It is important to note that seed emergence is an important factor for the rapid establishment of seedlings in field conditions.

Table 6. Accelerated ageing of sunflower seeds (%) versus storage time and type of packaging for each of the temperature and drying air storage conditions.

Drying air temperature	Conditions of storage	Packaging	Storage time (months) 0	3	6
		Waterproof	92 Aa	90 Aa	34 Ab
	20 °C / 60%	Permeable	92 Aa	86 Ba	36 Ab
45 °C		Waterproof	92 Aa	76 Bb	39 Ab
	25 °C / 50%	Permeable	92 Aa	96 Aa	32 Bb

		Waterproof	92 Aa	78 Ab	7 Ac
	30 °C / 40%	Permeable	92 Aa	0 Bb	0 Bb
		Waterproof	96 Aa	82 Bb	32 Ac
	20 °C / 60%	Permeable	96 Aa	86 Ab	31 Ac
55 °C		Waterproof	96 Aa	64 Bb	37 Ac
	25 °C / 50%	Permeable	96 Aa	82 Ab	31 Bc
		Waterproof	96 Aa	96 Aa	0 Ab
	30 °C / 40%	Permeable	96 Aa	0 Bb	0 Ab
		Waterproof	92 Aa	52 Ab	37 Ac
	20 °C / 60%	Permeable	92 Aa	68 Ab	30 Bc
65 °C		Waterproof	92 Aa	70 Bb	36 Ac
	25 °C / 50%	Permeable	92 Aa	92 Aa	33 Bb
		Waterproof	92 Aa	72 Ab	35 Ac
	30 °C / 40%	Permeable	92 Aa	0 Bb	0 Bb
		Waterproof	98 Aa	70 Ab	36 Ac
	20 °C / 60%	Permeable	98 Aa	68 Ab	29 Bc
75 °C		Waterproof	98 Aa	60 Bb	34 Ac
	25 °C / 50%	Permeable	98 Aa	90 Ab	0 Ac
		Waterproof	98 Aa	82 Ab	42 Ac
	30 °C / 40%	Permeable	98 Aa	0 Bb	0 B b

Averages followed by a capital letter in the column for each storage condition and a small letter in the rows for each drying air temperature do not differ at 5% probability. Coefficient of variation (CV) = 8.10%.

Seedlings with a higher EVI have a better performance and, consequently, a greater ability to withstand stresses that may interfere with the plant's growth and development. As well as the germination percentage, the germination speed index (GSE) was higher in seeds stored in environments with lower temperatures, as they reduced the respiration rate and maintained their quality for longer, compared to the other storage conditions. According to Guimaraes et al. (2005), bullgrass seeds had reduced EVI when stored without temperature and relative humidity control.

Table 7. Emergence rate of sunflower seeds (%) as a function of air drying temperature and storage conditions for each type of packaging and storage time.

Storage time (months)	Packaging	Storage conditions	Drying air temperature			
			45 °C	55 °C	65 °C	75 °C
0	-	-	2,07 c	2,36 b	1.49 d	2.03 a
3	Waterproof	20 °C / 60%	0.79 Ab	0.78 Ab	0.34 Bc	1.86 Ba
		25 °C / 50%	0.81 Ac	0.79 Ac	2.45 Aa	2.04 Ab
		30 °C / 40%	0Ba	0Ba	0Ca	0Ca
	Permeable	20 °C / 60%	0.81 Ba	0.81 Ba	0.82 Ba	0.78 Ba
		25 °C / 50%	2.36 Aa	2.39 Aa	2.36 Aa	2.46 Aa
		30 °C / 40%	0Ca	0Ca	0Ca	0Ca
6	Waterproof	20 °C / 60%	0.80 Aa	0.81 Aa	0.81 Ba	0.77 Bb
		25 °C / 50%	0.75 Bc	0.65 Bd	1.45 Ab	1.90 Aa
		30 °C / 40%	0Ca	0Ca	0Ca	0Ca
	Permeable	20 °C / 60%	0.65 Ba	0.60 Ba	0.54 Bb	0.48 Bc
		25 °C / 50%	1.82 Aa	1.71 Ab	1.16 Ac	1.10 Ad
		30 °C / 40%	0Ca	0Ca	0Ca	0Ca

Means followed by a capital letter in the column for each storage condition and a small letter in the rows for each drying air temperature do not differ at 5% probability. Coefficient of variation (CV) = 6.80%.

In general, there was a reduction in the emergence speed index (ESI) as the storage period increased, occurring mainly under conditions of 30 °C and 40% RH and permeable packaging (Tables 8 and 9). As for seedling growth (Figure 10), reductions were observed during the storage period, however, these reductions were more evident under storage conditions of 30 °C and 40% RH. Seed performance, as measured by the length of the aerial part and root system, was significantly different depending on storage time and packaging. Seeds stored for three months were influenced by the type of packaging used (Table 12 and 13).

Table 8. emergence rate of sunflower seeds (%) as a function of drying air temperature and storage time, for each type of packaging and storage condition

Storage time (months)	Packaging	Conditions of storage	Drying air temperature 45 °C	55 °C	65 °C	75 °C
0		-	2.07 Ab	2.36 Aa	1.49 Ac	2,03 Ab
3	Waterproof	20 °C / 60%	0.79 Bb	0.78 Bb	0.34 Cc	1,86 Ba
6		20 °C / 60%	0.80 Ba	0.81 Ba	0.81 Ba	0,77 Ca
0		-	2.07 Ac	2.36 Ab	1.49 Ad	2,03 Aa
3	Permeable	20 °C / 60%	0.81 Ba	0.81 Ba	0.82 Ba	0,78 Ba
6		20 °C / 60%	0.65 Ca	0.60 Cb	0.54 Cc	0,48 Cc
0		-	2.07 Ab	2.36 Aa	1.49 Bc	2,03 Ab
3	Waterproof	25 °C / 50%	0.81 Bc	0.79 Bc	2.45 Aa	2,04 Ab
6		25 °C / 50%	0.75 Cc	0.65 Cd	1.45 Bb	1,90 Ba
0		-	2.07 Bb	2.36 Aa	1.49 Bc	2,03 Cb
3	Permeable	25 °C / 50%	2.36 Ab	2.39 Ab	2.36 Ab	2,46 Aa
6		25 °C / 50%	1.82 Ca	1.71 Ba	1.16 Cb	1,10 Bb
0		-	2.07 Ab	2.36 Aa	1.49 Ac	2,03 Ab
3	Waterproof	30 °C / 40%	0Ba	0Ba	0Ba	0 Ba
6		30 °C / 40%	0Ba	0Ba	0Ba	0 Ba
0		-	2.07 Ab	2.36 Aa	1.49 Ac	2,03 Ab
3	Permeable	30 °C / 40%	0Ba	0Ba	0Ba	0 Ba
6		30 °C / 40%	0Ba	0Ba	0Ba	0 Ba

Averages followed by a capital letter in the column for each storage condition and a small letter in the rows for each drying air temperature do not differ at 5% probability. Coefficient of variation (CV) = 5.34%.

Table 9. Sunflower seed emergence rate (%) versus storage time and type of packaging, for each air temperature drying and storage condition.

Drying air temperature	Storage conditions	Packaging	Storage time (months)		
			0	3	6
45 °C	20 °C / 60%	Waterproof	2.07 Aa	0.79 Ab	0.80 Ab
		Permeable	2.07 Aa	0.81 Ab	0.65 Bc
	25 °C / 50%	Waterproof	2.07 Aa	0.81 Bb	0.75 Bc
		Permeable	2.07 Ab	2.36 Aa	1.82 Ac
	30 °C / 40%	Waterproof	2.07 Aa	0 Ab	0 Ab
		Permeable	2.07 Aa	0Ab	0Ab
55 °C	20 °C / 60%	Waterproof	2.36 Aa	0.78 Ab	0.81 Ab
		Permeable	2.36 Aa	0.81 Ab	0.60 Bc
	25 °C / 50%	Waterproof	2.36 Aa	0.79 Bb	0.65 Bc
		Permeable	2.36 Aa	2.39 Aa	1.71 Ab
	30 °C / 40%	Waterproof	2.36 Aa	0 Ab	0 Ab
		Permeable	2.36 Aa	0Ab	0Ab
65 °C	20 °C / 60%	Waterproof	1.49 Aa	0.34 Bc	0.81 Ab
		Permeable	1.49 Aa	0.82 Ab	0.54 Bc
	25 °C / 50%	Waterproof	1.49 Ab	2.45 Aa	1.45 Ab
		Permeable	1.49 Ab	2.36 Ba	1.16 Bc
	30 °C / 40%	Waterproof	1.49 Aa	0 Ab	0 Ab
		Permeable	1.49 Aa	0Ab	0Ab
75 °C	20 °C / 60%	Waterproof	2.03 Aa	1.86 Ab	0.77 Ac
		Permeable	2.03 Aa	0.78 Bb	0.48 Bc
	25 °C / 50%	Waterproof	2.03 Aa	2.04 Ba	1.90 Ab
		Permeable	2.03 Ab	2.46 Aa	1.10 Bc
	30 °C / 40%	Waterproof	2.03 Aa	0 Ab	0 Ab
		Permeable	2.03 Aa	0Ab	0Ab

Averages followed by a capital letter in the column for each storage condition and a small letter in the rows for each drying air temperature do not differ at 5% probability. Coefficient of variation (CV) = 7.10%.

Table 10. Sunflower seedling growth (%) according to drying air temperature and storage conditions, for each type of packaging and storage time.

Time of storage	Packaging	Storage conditions	Drying air temperature			
			45 °C	55 °C	65 °C	75 °C
0	-	-	7,3 a	5,8 c	6,6 b	7,0 a
3	Waterproof	20 °C / 60%	5.2 Bc	4.6 Bd	5.8 Bb	6.2 Ba
		25 °C / 50%	7.1 Aa	5.6 Ac	6.4 Ab	6.8 Ab
		30 °C / 40%	0.0 Ca	0.0 Ca	0.0 Ca	0.0 Ca
	Permeable	20 °C / 60%	5.3 Ab	4.8 Bc	5.3 Bb	6.2 Ba
		25 °C / 50%	5.4 Ad	7.4 Aa	5.9 Ac	6.5 Ab
		30 °C / 40%	0.0 Ba	0.0 Ca	0.0 Ca	0.0 Ca
6	Waterproof	20 °C / 60%	4.8 Bc	4.6 Bc	5.2 Bb	5.7 Ba
		25 °C / 50%	6.5 Aa	4.9 Ac	5.7 Ab	6.3 Aa
		30 °C / 40%	0.0 Ca	0.0 Ca	0.0 Ca	0.0 Ca
	Permeable	20 °C / 60%	4.8 Ab	4.1 Bc	4.4 Bc	5.7 Aa
		25 °C / 50%	4.9 Ab	6.3 Aa	5.1 Ab	4.9 Bb
		30 °C / 40%	0.0 Ba	0.0 Ca	0.0 Ca	0.0 Ca

Averages followed by a capital letter in the column for each storage condition and a small letter in the rows for each drying air temperature do not differ at 5% probability. Coefficient of variation (CV) = 4.90%.

The process of seed deterioration was influenced by seed performance, so that an increase in seed deterioration leads to lower seedling and root growth, triggered by a loss of vigor, culminating in a loss of germination (SILVA et al., 2010). The loss of seed germination and the early detection of physiological deterioration are important aspects to consider in storage, so the percentage of humidity and storage temperature are the main factors that can be manipulated and controlled to minimize the deterioration of seeds preserved in the long term (ALMEIDA et al., 2010).

Table 11. Root growth in sunflower seeds (%) according to drying air temperature and storage conditions, for each type of packaging and storage time.

Storage time (months)	Packaging	Storage conditions	Drying air temperature			
			45 °C	55 °C	65 °C	75 °C
0	-	-	5,2 c	5,5 c	6.8 b	7,4 a
3	Waterproof	20 °C / 60%	4.1 Ba	4.1 Ba	3.6 Bb	3.5 Bb
		25 °C / 50%	6.5 Aa	4.9 Ab	6.3 Aa	6.6 Aa
		30 °C / 40%	0.0 Ca	0.0 Ca	0.0 Ca	0.0 Ca
	Permeable	20 °C / 60%	3.0 Bc	2.7 Bc	5.8 Bb	7.0 Aa
		25 °C / 50%	3.8 Ac	4.0 Ac	7.0 Aa	4.9 Bb
		30 °C / 40%	0.0 Ca	0.0 Ca	0.0 Ca	0.0 Ca
6	Waterproof	20 °C / 60%	3.8 Ba	3.2 Bb	2.8 Bb	3.0 Bb
		25 °C / 50%	5.7 Ab	4.0 Ac	5.7 Ab	6.0 Aa
		30 °C / 40%	0.0 Ca	0.0 Ca	0.0 Ca	0.0 Ca
	Permeable	20 °C / 60%	2.5 Bc	2.2 Bc	4.5 Bb	6.2 Aa
		25 °C / 50%	2.8 Ad	3.4 Ac	5.5 Aa	4.3 Bb
		30 °C / 40%	0.0 Ca	0.0 Ca	0.0 Ca	0.0 Ca

Averages followed by a capital letter in the column for each storage condition and a small letter in the rows for each drying air temperature do not differ at 5% probability. Coefficient of variation (CV) = 3.85%.

Table 12. Sunflower seedling growth (%) versus storage time and type of packaging, for each of the temperature and drying air storage conditions.

Drying air temperature	Storage conditions	Packaging	Storage time (months)		
			0	3	6
45 °C	20 °C / 60%	Waterproof	7.3 Aa	5.2 Ab	4.8 Ac
		Permeable	7.3 Aa	5.3 Ab	4.8 Ac
		Waterproof	7.3 Aa	7.1 Aa	6.5 Ab

	25 °C / 50%	Permeable	7.3 Aa	5.4 Bb	4.9 Bc
		Waterproof	7.3 Aa	0.0 Ab	0.0 Ab
	30 °C / 40%	Permeable	7.3 Aa	0.0 Ab	0.0 Ab
		Waterproof	5.8 Aa	4.6 Ab	4.6 Ab
	20 °C / 60%	Permeable	5.8 Aa	4.8 Ab	4.1 Bc
55 °C		Waterproof	5.8 Aa	5.6 Ba	4.9 Bb
	25 °C / 50%	Permeable	5.8 Ac	7.4 Ba	6.3 Ab
		Waterproof	5.8 Aa	0.0 Ab	0.0 Ab
	30 °C / 40%	Permeable	5.8 Aa	0.0 Ab	0.0 Ab
		Waterproof	6.6 Aa	5.8 Ab	5.2 Ac
	20 °C / 60%	Permeable	6.6 Aa	5.3 Bb	4.4 Bc
65 °C		Waterproof	6.6 Aa	6.4 Aa	5.7 Ab
	25 °C / 50%	Permeable	6.6 Aa	5.9 Bb	5.1 Bc
		Waterproof	6.6 Aa	0.0 Ab	0.0 Ab
	30 °C / 40%	Permeable	6.6 Aa	0.0 Ab	0.0 Ab
		Waterproof	7.0 Aa	6.2 Ab	5.7 Ac
	20 °C / 60%	Permeable	7.0 Aa	6.2 Ab	5.7 Ac
75 °C		Waterproof	7.0 Aa	6.8 Aa	6.3 Ac
	25 °C / 50%	Permeable	7.0 Aa	6.5 Bb	4.9 Bc
		Waterproof	7.0 Aa	0.0 Ab	0.0 Ab
	30 °C / 40%	Permeable	7.0 Aa	0.0 Ab	0.0 Ab

Averages followed by a capital letter in the column for each storage condition and lower case on the lines for each drying air temperature do not differ at 5% probability. Coefficient of variation (CV) = 7.22%.

Table 13. Root growth in sunflower seeds (%) versus storage time and type of packaging, for each of the drying air temperature and storage conditions.

Drying air temperature	Storage conditions	Packaging	Storage time (months)		
			0	3	6
45 °C	20 °C / 60%	Waterproof	5.2 Aa	4.1 Ab	3.8 Ac
		Permeable	5.2 Aa	3.0 Bb	2.5 Bc
	25 °C / 50%	Waterproof	5.2 Ac	6.5 Aa	5.7 Ab
		Permeable	5.2 Aa	3.8 Bb	2.8 Bc
	30 °C / 40%	Waterproof	5.2 Aa	0.0 Ab	0.0 Ab
		Permeable	5.2 Aa	0.0 Ab	0.0 Ab
55 °C	20 °C / 60%	Waterproof	5.5 Aa	4.1 Ab	3.2 Ac
		Permeable	5.5 Aa	2.7 Bb	2.2 Bc
	25 °C / 50%	Waterproof	5.5 Aa	4.9 Ab	4.0 Ac
		Permeable	5.5 Aa	4.0 Bb	3.4 Bc
	30 °C / 40%	Waterproof	5.5 Aa	0.0 Ab	0.0 Ab
		Permeable	5.5 Aa	0.0 Bb	0.0 Ab
65 °C	20 °C / 60%	Waterproof	6.8 Aa	3.6 Bb	2.8 Bc
		Permeable	6.8 Aa	5.8 Ab	4.5 Ac
	25 °C / 50%	Waterproof	6.8 Aa	6.3 Bb	5.7 Ac
		Permeable	6.8 Aa	7.0 Aa	5.5 Ab
	30 °C / 40%	Waterproof	6.8 Aa	0.0 Ab	0.0 Ab
		Permeable	6.8 Aa	0.0 Ab	0.0 Ab
75 °C	20 °C / 60%	Waterproof	7.4 Aa	3.5 Bb	3.0 Bc
		Permeable	7.4 Aa	7.0 Ab	6.2 Ac
	25 °C / 50%	Waterproof	7.4 Aa	6.6 Ab	6.0 Ac
		Permeable	7.4 Aa	4.9 Bb	4.3 Bc
	30 °C / 40%	Waterproof	7.4 Aa	0.0 Ab	0.0 Ab
		Permeable	7.4 Aa	0.0 Ab	0.0 Ab

Averages followed by a capital letter in the column for each storage condition and lower case on the lines for each drying air temperature do not differ at 5% probability. Coefficient of variation (CV) = 6.25%.

CONCLUSIONS

The physiological potential of sunflower seeds was affected by drying temperatures and storage conditions. Drying air temperatures above 45 °C negatively affected the quality of sunflower seeds. Permeable and impermeable packaging does not guarantee the quality of sunflower seeds during storage. The quality of sunflower seeds stored for three months was affected. The conditions of 20 °C and 60% RH were more favorable for preserving the physiological quality of sunflower seeds during the storage period.

BIBLIOGRAPHICAL REFERENCES

ALMEIDA, F. A. C. et al. Study of techniques for the storage of five oilseeds in ambient and cryogenic conditions. **Revista Brasileira de Produtos Agroindustriais**, Campina Grande, v. 12, n. 2, p. 189-202.

AZEVEDO, M. R. Q. A. et al. Influence of packaging and storage conditions on the vigor of sesame seeds. **Revista Brasileira de Engenharia Agricola e Ambiental**, v. 7, n. 3, p. 519-524, 2003.

BAILEY, J. E. Whole grain storage. In: SAUER, D.B. (Ed.). Storage of cereal grains and their products. Minnesota: **American Association of Cereal Chemists**, 1992. 615p.

BENNETT, M. A. **Seed vigor and seed performance**. Available at: <http://www.ag.ohiostate.edu/seedbio/hcs631.htm>. Accessed on: 22 Jan. 2011.

BRAZIL. Ministry of Agriculture, Livestock and Supply. **Rules for seed analysis**. Brasilia: SNDA/DNDV/CLAV, 2009, 365p.

CARVALHO, N. M.; NAKAGAWA, J. **Seeds:** science, technology and production**.** 4. ed. Jaboticabal: Funep, 2000.

CLEMENTE, A. da C. S. et al. Electrical conductivity test on soybean seeds stored at different temperatures. **Bragantia**, v. 69, n. 1, p. 207-214, 2010.

GUIMARAES, R. N. et al. Effect of Seed Treatment with Insecticides on the Emergence and Height of Bean Seedlings. In: CONGRESSO NACIONAL DE PESQUISA DE FEIJAO, 8, 2005, Goiânia. **Proceedings...** Goiânia: Embrapa Arroz e Feijao, 2005, p. 94-99.

GUIMARAES, R. M.; ZEVIANI, W. M. Preparation of coffee seeds for viability assessment by the tetrazolium test. **Revista Brasileira de Sementes**, v. 33, p. 38-44, 2011. MAGUIRE, J. D. Speed of germination - aid in selection and evaluation for seedling emergence and vigor. **Crop Science**, v. 2, n. 2, p. 176-177, 1962.

MARCOS FILHO, J. **Fisiologia de sementes de plantas cultivadas**. Piracicaba: Fealq, 2005. 495p.

MARTINS, L.; LAGO, A. A.; ANDRADE, A. C. S. Storage of ipê branco seeds: water content and ambient temperature. **Bragantia**, v. 68, n. 3, p. 775-780, 2009.

PRADHAN, R. C. et al. Moisture-dependent physical properties of jatropha fruit. **Industrial Crops and Products,** Amsterdam, v. 29, n. 2, p. 341-347, 2009.

NAKAGAWA, J. Vigor tests based on seedling performance. In: KRZYZANOWSKI, F.C. **Seed vigor**: concepts and tests. Londrina: ABRATES, 1999, p. 21-24.

SANTOS, J. F. et al. Optimizing tetrazolium test procedures to evaluate the physiological potential of peanut seeds **Seed Science and Technology**, v. 40, p. 215-228, 2012.

SILVA, F. S. et al. Feasibility of seed storage in different packages for small rural properties. **Revista de Ciências Agro-Ambientais**, Alta Floresta, v. 8, n. 1, p. 45- 56, 2010.

TOLEDO, M. Z. et al. Physiological quality and storage of bean seeds as a function of late nitrogen top dressing. **Pesquisa Agropecuària Tropical**, Goiânia, v. 39, n. 2, p. 124-133, 2009.

TONIN, G. A.; PEREZ, S. C. J. G. A. Physiological quality of *Ocotea porosa* (Nees et Martius ex. Nees) seeds after different storage and sowing conditions. **Revista Brasileira de Sementes**, Pelotas, v. 28, n. 2, p. 26-33, 2006.

MIX
Papier aus verantwortungsvollen Quellen
Paper from responsible sources
FSC® C105338

Printed by Books on Demand GmbH, Norderstedt / Germany